WORLD ABOUT US

THE OZONE LAYER

M. BRIGHT

GLOUCESTER PRESS
New York London Toronto Sydney

© Aladdin Books 1991

First published in
the United States in 1991 by
Gloucester Press
387 Park Avenue South
New York, NY 10016

Design: David West
Children's
Book Design
Editor: Fiona Robertson
Illustrator: James Macdonald
Consultant: Brian Gardiner

Library of Congress
Cataloging-in-Publication Data

Bright, Michael.
The ozone layer / Michael Bright.
p.   cm. -- (World about us)
Includes index.
Summary: Explains the nature
and importance of the ozone
layer, how it is most likely
damaged, and what measures
are being taken to protect it.
ISBN 0-531-17302-X
1. Chlorofluorocarbons-
-Environmental apects-
-Juvenile literature. 2. Ozone
layer depletion--Environmental
aspects--Juvenile literature.
[1. Ozone layer.] I. Title.
II. Series: Bright, Michael. World
about us.
TD887.C47B75   1991
363.73'84--dc20
90-45658   CIP   AC

Contents

Dangers and benefits
4
Protecting earth
6
The Ozone Layer
8
"Ozone-eaters"
10
CFC sources
12
The effects
14
Sunbathers beware
16
Down here
19
Ozone hole
20
Rich and poor
22
Alternatives
25
The green issues
26
Ozone facts
28
Glossary
31
Index
32

Introduction

The liquid used to cool the air in a refrigerator could be harmful to life on earth. When old refrigerators are destroyed, harmful gases can escape into the atmosphere. The atmosphere is the blanket of gases that surrounds the earth. It is made up of several layers. One of these layers contains ozone, a gas that protects us from harmful ultraviolet light.

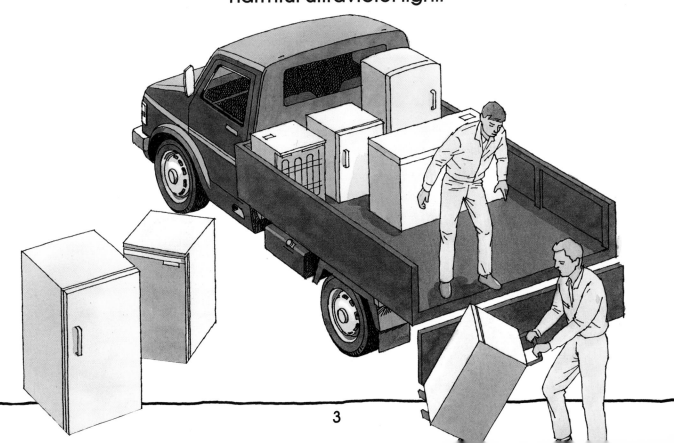

Dangers and benefits

The sun and the atmosphere make life on earth possible. The earth is kept warm by the sun's heat, and the atmosphere traps some of this heat so that it doesn't escape into space. But not all the energy made by the sun is safe. Dangerous forms of radiation called ultraviolet, or UV, light are also given out, and these can be harmful to life.

A rainbow is made up of seven different colors: red, orange, yellow, green, blue, indigo, and violet.

Heat and light from the sun allow plants to grow.

4

The sun is just the right distance away from earth to warm us.

Animals cannot use the energy from the sun directly, and so have to depend on plants for food.

Protecting earth

The atmosphere is like an invisible shield of air surrounding the earth. It contains different gases which protect life on the planet. The atmosphere lets useful energy through, but reduces the amount of harmful energy reaching the earth's surface. At a height of six to 18 miles above us, there is a layer in the atmosphere containing ozone. This stops some of the harmful UV light getting to earth.

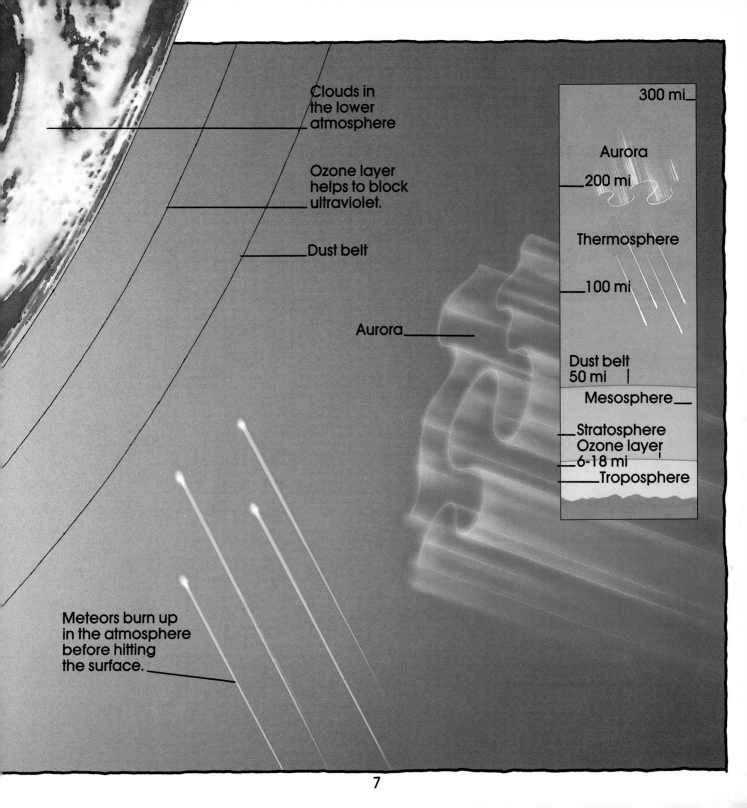

Clouds in
the lower
atmosphere

Ozone layer
helps to block
ultraviolet.

Dust belt

Aurora

Meteors burn up
in the atmosphere
before hitting
the surface.

300 mi

Aurora

200 mi

Thermosphere

100 mi

Dust belt
50 mi

Mesosphere

Stratosphere
Ozone layer
6-18 mi

Troposphere

The Ozone Layer

Ozone is a form of oxygen. It is made when ultraviolet radiation from the sun meets oxygen in the atmosphere. The ozone layer is found in the stratosphere (see diagram page 7). It stops most of the dangerous ultraviolet radiation from the sun getting to earth. However, ozone can be easily destroyed in the atmosphere. This means that we are putting the ozone layer under threat.

Measuring ozone
There are many different ways of measuring the amount of ozone in the atmosphere. Scientists use high-flying aircraft and satellites to gather their information. In addition, weather balloons (shown right) carry measuring instruments up into the sky. Their findings are then sent back to earth.

Information gathered by weather balloons can tell us more about the ozone layer.

9

"Ozone-eaters"

The gases used in refrigerators are chemicals called chlorofluorocarbons (CFCs). They are safe at ground level, but are dangerous in the atmosphere.

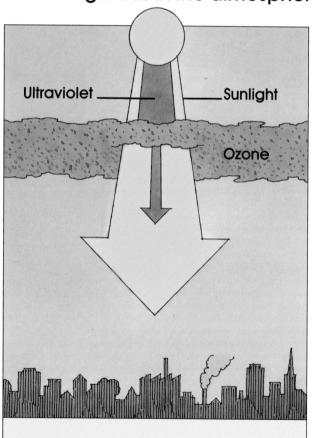

Ultraviolet — Sunlight

Ozone

The ozone layer reduces the amount of ultraviolet reaching the earth from the sun's rays.

CFCs are also found in some aerosols. They can stay in the atmosphere for a very long time.

Once CFCs are in the stratosphere they can last for 100 years or more before they are destroyed. Here they are broken down into chemicals by ultraviolet light from the sun. One of these chemicals is chlorine, which destroys the ozone layer.

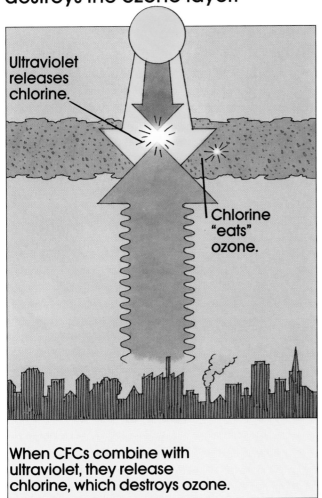

Ultraviolet releases chlorine.

Chlorine "eats" ozone.

When CFCs combine with ultraviolet, they release chlorine, which destroys ozone.

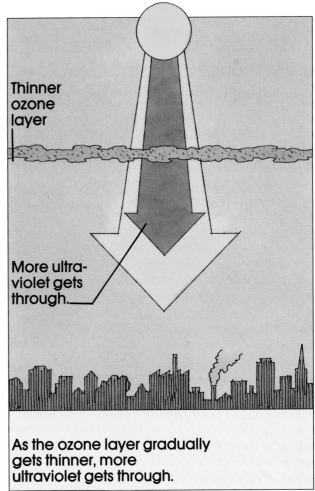

Thinner ozone layer

More ultraviolet gets through.

As the ozone layer gradually gets thinner, more ultraviolet gets through.

CFC sources

CFCs are also found in freezers and air conditioners, and in some aerosol sprays and fire extinguishers. Plastic foams, used for packaging and furniture, can be made with CFCs. In electronics factories, CFCs are used to clean the circuit boards that go into televisions and computers. The CFCs then gradually drift up into the atmosphere.

Some companies use CFCs to expand the foam packaging for burgers and eggs.

Some glues, pens and paints use chemicals that can destroy the ozone layer.

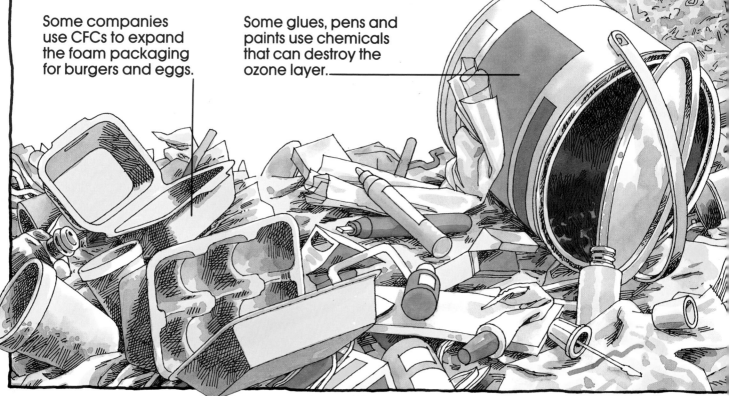

The liquid used to cool refrigerators and air conditioners contains CFCs.

Nitrogen oxides in exhaust fumes of high-flying aircraft may harm the ozone layer.

Gases in some fire extinguishers damage ozone.

The effects

If we allowed the ozone layer to get too thin, large amounts of harmful ultraviolet light would reach the earth's surface. Seeds would not grow, flowering would stop, and many plants would die. Without plants, humans and animals would soon starve. Even a little damage to the ozone layer could harm our food crops.

Plankton in the sea will suffer. This is the lowest part of the food chain, which means that everything else will be affected.

Many plants, including our crops, cannot flower or produce seeds. The crops will die.

Sunbathers beware

The body is protected from ultraviolet light by special substances in the skin, called pigments. It is these pigments that make white skin turn brown, or tan, after sunbathing. If the ozone layer gets thinner and lets in too much ultraviolet light, the body's protection system becomes less effective. Then the skin can be badly damaged, causing skin cancer. Damage to the eyes can cause blindness.

Lots of factories, cars and buses often make pollution much worse in big cities.

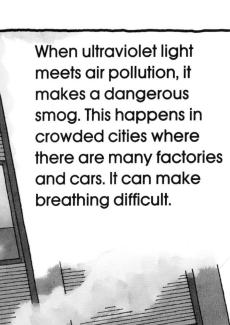

When ultraviolet light meets air pollution, it makes a dangerous smog. This happens in crowded cities where there are many factories and cars. It can make breathing difficult.

Down here

High up in the sky, the ozone layer protects the earth from the sun's most harmful rays. But at ground level, ozone is a dangerous form of pollution, that can be harmful to life. Ozone is formed in cities when the unburned fuel from car exhausts mixes with sunlight. If there is not enough wind to carry the ozone up into the atmosphere, it builds up near the ground. It forms a dangerous smog which can cause people to suffer from health problems.

People who work outside, like the police or bicycle messengers, may have to wear face masks.

Car exhausts react with ultra-violet light to form a smog which contains ozone.

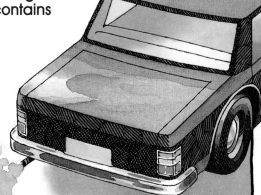

Ozone hole

Scientists have been measuring ozone in the Antarctic for about 30 years. In the 1980s, they noticed that a large area of the ozone layer over the Antarctic was becoming very thin each spring. Nearly all the ozone there has been destroyed by CFCs. In the 1960s, there was no ozone hole because CFCs were hardly used. Now there is an ozone hole over the Antarctic every spring. It will keep on happening until we stop using CFCs.

If we keep putting more and more CFCs into the atmosphere, the ozone will continue to be destroyed.

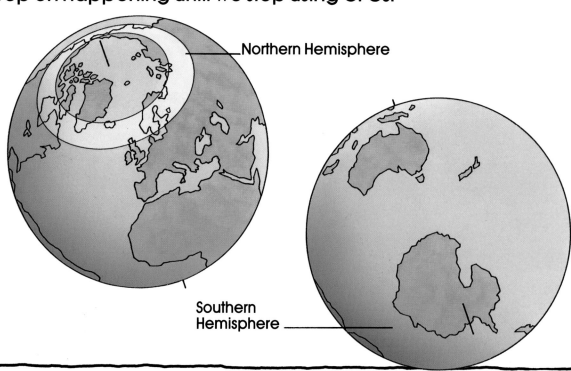

Northern Hemisphere

Southern Hemisphere

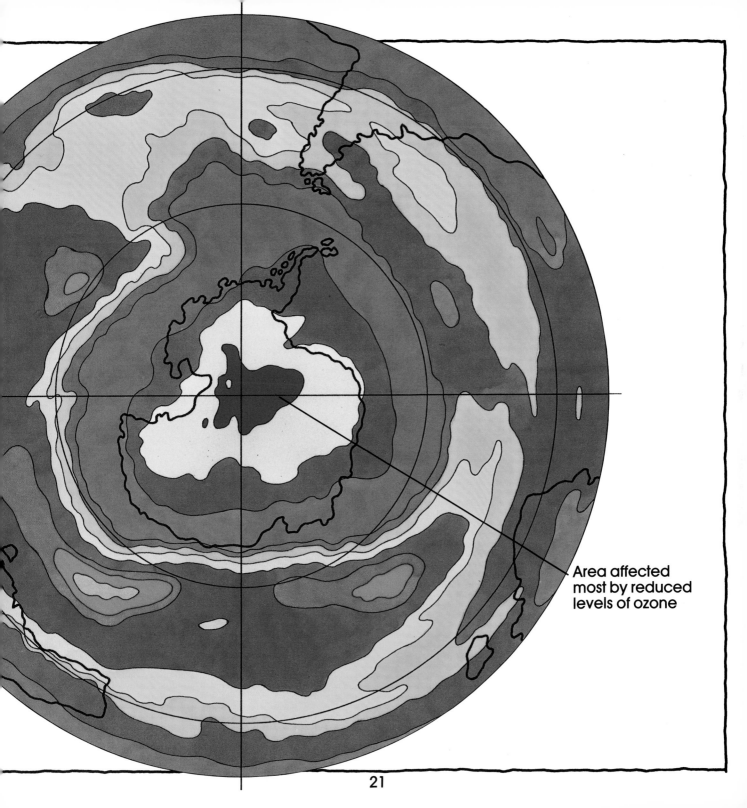

Area affected
most by reduced
levels of ozone

Rich and poor

In many developed countries, new laws have been passed to cut down on the use of CFCs. Cars can be specially designed to reduce the amount of harmful fumes they give out into the atmosphere. However, in many developing countries, it is harder to ban the use of CFCs. Aerosols and refrigerators are seen as luxuries, and products without CFCs are often expensive.

Some cars have air conditioners that use CFCs. In wealthier countries, older cars can be towed away, and the CFCs then safely recycled.

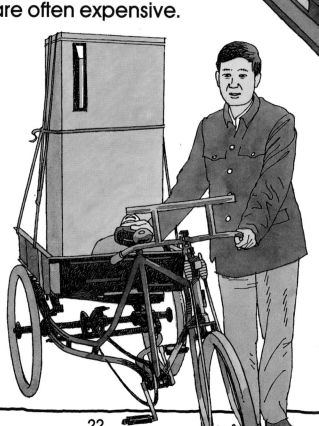

Some poorer countries are not wealthy enough to recycle old CFCs or use alternative products. This is a problem.

Companies
can come and
collect old
refrigerators
and recycle
the CFCs.

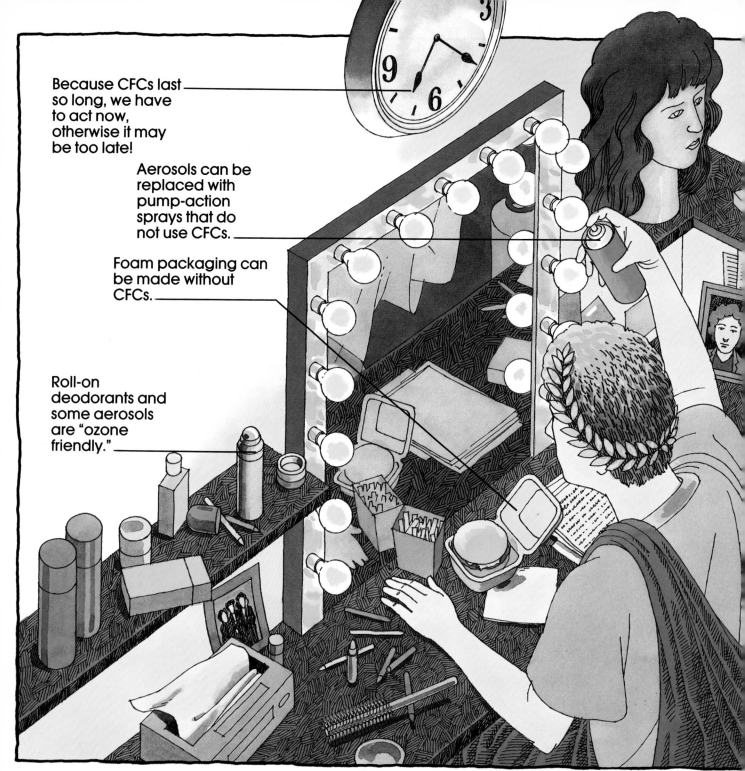

Because CFCs last so long, we have to act now, otherwise it may be too late!

Aerosols can be replaced with pump-action sprays that do not use CFCs.

Foam packaging can be made without CFCs.

Roll-on deodorants and some aerosols are "ozone friendly."

Alternatives

Many of the CFCs used today could be replaced with other, safer products. For example, pump-action sprays that do not use harmful gases could be used instead of aerosols. In the future, refrigerators will use new gases that do not damage the ozone layer. Foam packaging can be made without using CFCs.

Carbon dioxide is now used in foam fire extinguishers. But carbon dioxide is a greenhouse gas. This means that it helps to trap heat in the atmosphere. If too much heat is trapped, temperatures may go up. This is known as global warming.

Some aerosols have symbols on the can (see right) to show that they are not harmful to the environment.

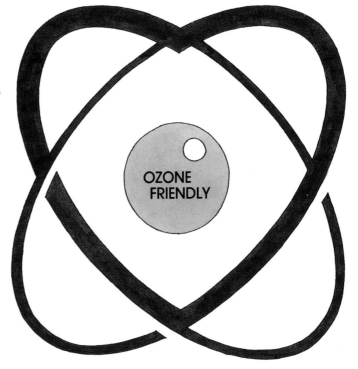

OZONE FRIENDLY

The green issues

Governments around the world became so concerned about the threat of CFCs to the ozone layer that they signed an agreement in 1987 called the Montreal Protocol. The countries that signed the agreement have promised to stop using the worst CFCs by the end of the century. But all the gases which could destroy the ozone layer should be banned, not just the worst ones.

China, India and some other developing countries have not yet signed the Montreal Protocol. They will only sign it if the developed countries will help them pay for CFC alternatives.

Ozone facts

When you release some CFC gas into the air, it travels all over the world for about 100 years. Eventually, it finds its way up into the stratosphere, where it is broken down by strong ultraviolet light. By that time, your great-grandchildren will have been to school.

All plants and food crops need sunlight to make them grow. If we damage the ozone layer, the plants cannot escape from the harmful ultraviolet sunlight. Fungi can grow in the dark forest, but only because they use the dead plants for food.

CFCs are thousands of times better at trapping heat in the atmosphere than carbon dioxide. So not only are they destroying the ozone layer, but they are also contributing to global warming. Banning CFCs will help to reduce this.

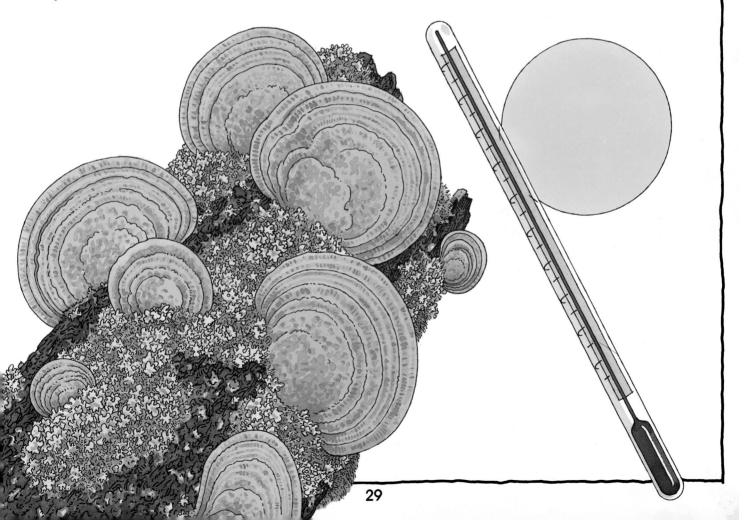

Everything is made up of atoms. When atoms join together they form molecules. A CFC molecule contains chlorine atoms. When the molecule breaks up, the chlorine atom is released. It can then easily destroy 100,000 ozone molecules. That is why we must stop using CFCs before any more of the ozone layer is destroyed.

Scientists in the Antarctic have been checking the amount of ozone in the atmosphere since the 1950s. At that time there was no such thing as an "ozone hole." But as we began to use more and more CFCs in the 1970s, the ozone layer started to disappear. Now CFCs destroy more than half of the ozone over Antarctica each spring.

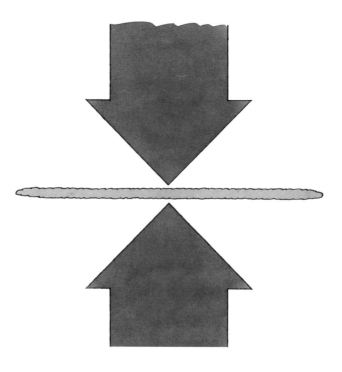

Glossary

Atmosphere
The layer of gases that surrounds the Earth. It is about 430mi thick.

CFCs / chlorofluorocarbons
Chemicals used in products like aerosols and refrigerators.

Global warming
Pollution has increased the Greenhouse Effect by putting more gases that trap heat into the atmosphere. This makes the earth's temperature go up. This is called global warming.

Greenhouse Effect
The normal process by which heat is kept in the atmosphere. Without it, life could not survive on earth. But if too much heat is trapped, temperatures could go up. This could be dangerous to life.

Montreal Protocol
An international agreement, signed by many countries, which aims to help save the ozone layer. The countries that signed have promised to take the action necessary to reduce pollution.

Ozone
A colorless gas which is a form of ozone.

Ozone layer
A band of ozone in the atmosphere which stops ultraviolet light getting to earth.

Ultraviolet light
Invisible light from the sun which makes skin go brown, or tan. But it can also cause skin cancers.

Index

A

aerosol cans 10, 12, 22, 24, 25, 28

alternatives to CFCs 25

atmosphere 4, 6-8, 10, 14, 19, 20, 30

C

carbon dioxide 14, 29

CFCs 10-13, 22-29

chlorine 11, 30

E

energy 4-6

F

foam packaging 24, 25, 28

G

Greenhouse Effect 29

M

Montreal Protocol 26

O

ozone 3, 6, 8, 11, 19

ozone hole 20, 30

ozone layer 8-11, 14, 16, 19, 20, 26, 29, 30

P

pigments 16

pollutants 8, 12, 13, 18, 22

pollution 18, 19

R

recycling 22, 23

refrigerators 3, 12, 13, 22, 23, 25

S

smog 19

sun 4, 5

sunbathing 16

U

ultraviolet light 3, 4, 6, 8, 10, 11, 14, 16, 19

PRINTED IN BELGIUM BY

proost

INTERNATIONAL BOOK PRODUCTION